BUILDING BLOCKS OF MATH

GEOMETRY

Written by Regina Osweiller

Illustrated by Daniel Hawkins

WORLD BOOK
a Scott Fetzer company
Chicago

World Book, Inc.
180 North LaSalle Street
Suite 900
Chicago, Illinois 60601
USA

For information about other World Book publications, visit our website at **www.worldbook.com** or call **1-800-WORLDBK (967-5325)**.
For information about sales to schools and libraries, call 1-800-975-3250 (United States), or 1-800-837-5365 (Canada).

WORLD BOOK STAFF
Executive Committee
President: Geoff Broderick
Vice President, Editorial: Tom Evans
Vice President, Finance: Donald D. Keller
Vice President, Marketing: Jean Lin
Vice President, International: Eddy Kisman
Vice President, Technology: Jason Dole
Director, Human Resources: Bev Ecker

Editorial
Manager, New Content: Jeff De La Rosa
Associate Manager, New Product: Nicholas Kilzer
Sr. Editor: William M. Harrod
Proofreader: Nathalie Strassheim

Graphics and Design
Sr. Visual Communications Designer: Melanie Bender
Sr. Web Designer/Digital Media Developer: Matt Carrington

Library of Congress Cataloging-in-Publication Data for this volume has been applied for.

Building Blocks of Math
ISBN: 978-0-7166-4253-4 (set, hc.)

Geometry
ISBN: 978-0-7166-4259-6 (hc.)

Also available as:
ISBN: 978-0-7166-4269-5 (e-book)

Printed in India by Thomson Press (India) Limited, Uttar Pradesh, India
1st printing June 2022

Acknowledgments:
Writer: Regina Osweiller
Illustrator: Daniel Hawkins/The Bright Agency
Series Advisor: Marjorie Frank
Special thanks to KnowledgeWorks Global Ltd.

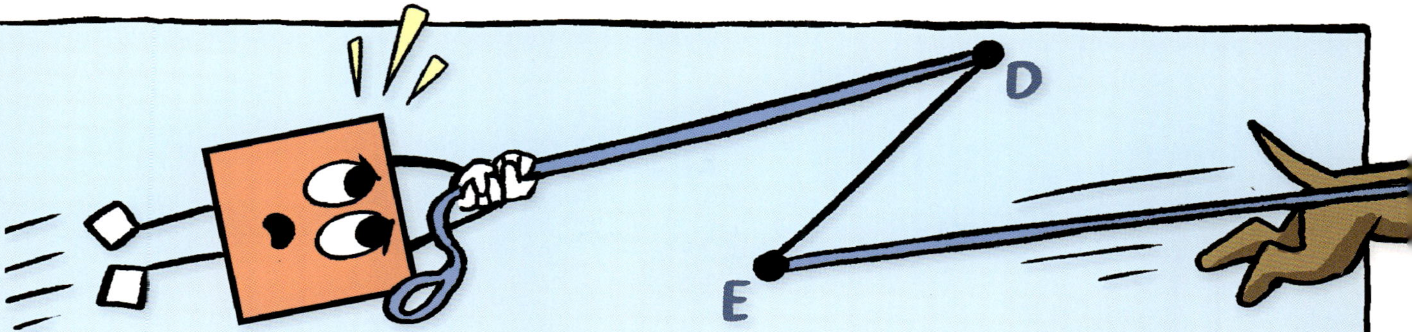

TABLE OF CONTENTS

GEOMETRY WORLD
Here is your world.
GEE I'M A TREE

And here is Geometry World, where I live with my friends.
3-D GEOMETRY WORLD
Geometry is the branch of math where we study points, lines, surfaces, and shapes. We make measurements and identify properties of these figures.
GEO-METRY

SNIFF SNIFF
CRASH!!
Hey! You can't come in here!
CHOMP
Oh no! The dog is ruining everything!
We will have to rebuild Geometry World.
3-D GEOMETRY WORLD

PLANES, POINTS, LINES, AND ANGLES

Let's add some points.
A point is an exact position on a plane. It has no length, width, or height. To show where a point is, we'll make it into a small dot.
B
A
Can I name this point C, for Chaos?
C
Sure. Points are named with letters.
D
E

A **line** is a set of points that extends forever at both ends. A line is usually named with two of its points and a double arrow, like this: $\overleftrightarrow{AB}$.
m
B
A
You can also name a line with a lowercase letter, like *m*.
D
E

What's this?
D
This figure is called a ray. A **ray** has a point at one end, called an **endpoint**. The ray goes on forever away from the endpoint.
E
F
A ray is named with its endpoint, another point on the ray, and a one-way arrow, like this: $\overrightarrow{EF}$.
I'm drawing a **line segment**, which is a part of a line that is between two endpoints.
My line segment is named $\overline{GH}$. Chaos is balancing on line segment $\overline{KL}$.
H
G
K
L

My goodness! Where did you come from?
You just made me! I'm Angle.
A
B
C
An **angle** forms from two line segments or rays that share a common endpoint. That endpoint is called the **vertex**.
Angles are also formed when two lines or planes intersect (cross).
An angle is named with an angle symbol and three of its points, with the vertex in the middle. Our new friend is named like this: ∠ABC.
A
C
B
A
C
B
I can be as narrow or as wide as I need to be.
A
120°
B
C
Use a **protractor** to measure an angle. **Degrees** are the units of angle measurement.

Looks like I have four angles. What are their measurements?
Each of your angles measures 90 degrees, which is called a **right angle**. We can show a right angle with a corner mark.
90°
A
C
B
An **acute angle** measures less than 90 degrees.
An **obtuse angle** measures between 90 degrees and 180 degrees.
A
B
C
And now I'll widen myself even more.
HNNNG!
This is not a corner mark, it's just the page number!
A
B
C

!?
What happened to Angle?
I'm down here!
I'm now a **straight angle**, which is an angle that measures 180 degrees.
You look like a straight line.
That's exactly right. A straight angle is formed by three points on the same line.
A
B
C

PARALLEL AND PERPENDICULAR LINES
m
n
CHIRP CHIRP
Most lines in a plane will **intersect**, or cross over, to form angles.
But do you see the two arrow marks on lines *m* and *n*? The marks show **parallel lines**, which are lines in the same plane that never intersect.

CRASH!!
Aha, lines m and n have become perpendicular lines.
90°
A
B
C
Perpendicular lines meet to form right angles (90-degree angles).

Hey, Triangle, what are you doing?
Well, a **triangle** always has three sides and three interior angles.
But otherwise, triangles can look different from one another.
Check this out!
Right Triangle
One angle is a right angle that measures 90°.
Obtuse Triangle
One angle is an obtuse angle that measures more than 90°.

Do you notice the hash marks on my sides? They show which sides have the same length.
THEATER
Equilateral Triangle
All sides have the same length. Each angle measures 60°.
Isosceles Triangle
Two sides have the same length, and their opposite angles have equal measures.
Scalene Triangle
All three sides and all three angles are different.
Professor Circle
60° + 60° + 60° = 180°
30° + 60° + 90° = 180°
Do you see a pattern in the sums of the angles of a triangle? Their measures always add up to 180°.

QUADRILATERALS
Triangles are great, but check out these quadrilaterals.
A **quadrilateral** is a two-dimensional figure with four sides and four interior angles. *Quadri-* means "four," and *lateral* means "sides."
Square
Four sides have equal length. There are four right angles.
Rhombus
This parallelogram has four sides of equal length.
Rectangle
This parallelogram has four right angles.
I'm a parallelogram, a rectangle, and a rhombus!

QUADRILATERALS
Parallelogram
Opposite sides are parallel and equal in length. Opposite angles are equal.
Trapezoid
Only one pair of sides is parallel.
CLICK

EXIT
Do you think that the four angles of a quadrilateral always have the same sum, like the angles of a triangle do?
I know a way to find out...
SLURP

Two triangles can always fit inside a quadrilateral. So, the sum of the angles of a quadrilateral is always twice 180°, or 360°.

Hmm, I've never been to Polygon Park before.
What's a polygon?
POLYGON PARK
Pentagon
five sides and five angles
STOP
Octagon
eight sides and eight angles

That's an interesting question because both of you are polygons.
We are?
Yes. A **polygon** is a two-dimensional figure with three or more straight sides. It also has the same number of angles as sides.
Hexagon six sides and six angles
Quadrilateral four sides and four angles
Triangle three sides and three angles

Hey, look at our polygons!
Ah, you are drawing **irregular polygons.** The sides have different lengths, and the angles have different measures.

I am drawing **regular polygons,** which have sides of the same length. All interior angles are the same, too.

Are you a polygon, Circle?
No, I am not a polygon. Polygons have angles and straight sides, and circles don't!
Drat, that's not the polygon I wanted.

Square, watch out! You're erasing the edge of the page!
So, what can happen?

THREE-DIMENSIONAL FIGURES
THAT'S what can happen! Not all of Geometry World is flat and two dimensional, you know.
THREE-DIMENSIONAL GEOMETRY WORLD
What are these things?
These are **three-dimensional figures**. They have length, width, and height.
Cube
6 square faces joined at right angles
6 square faces, 8 right-angle vertices, 12 edges
DIMENSIONAL

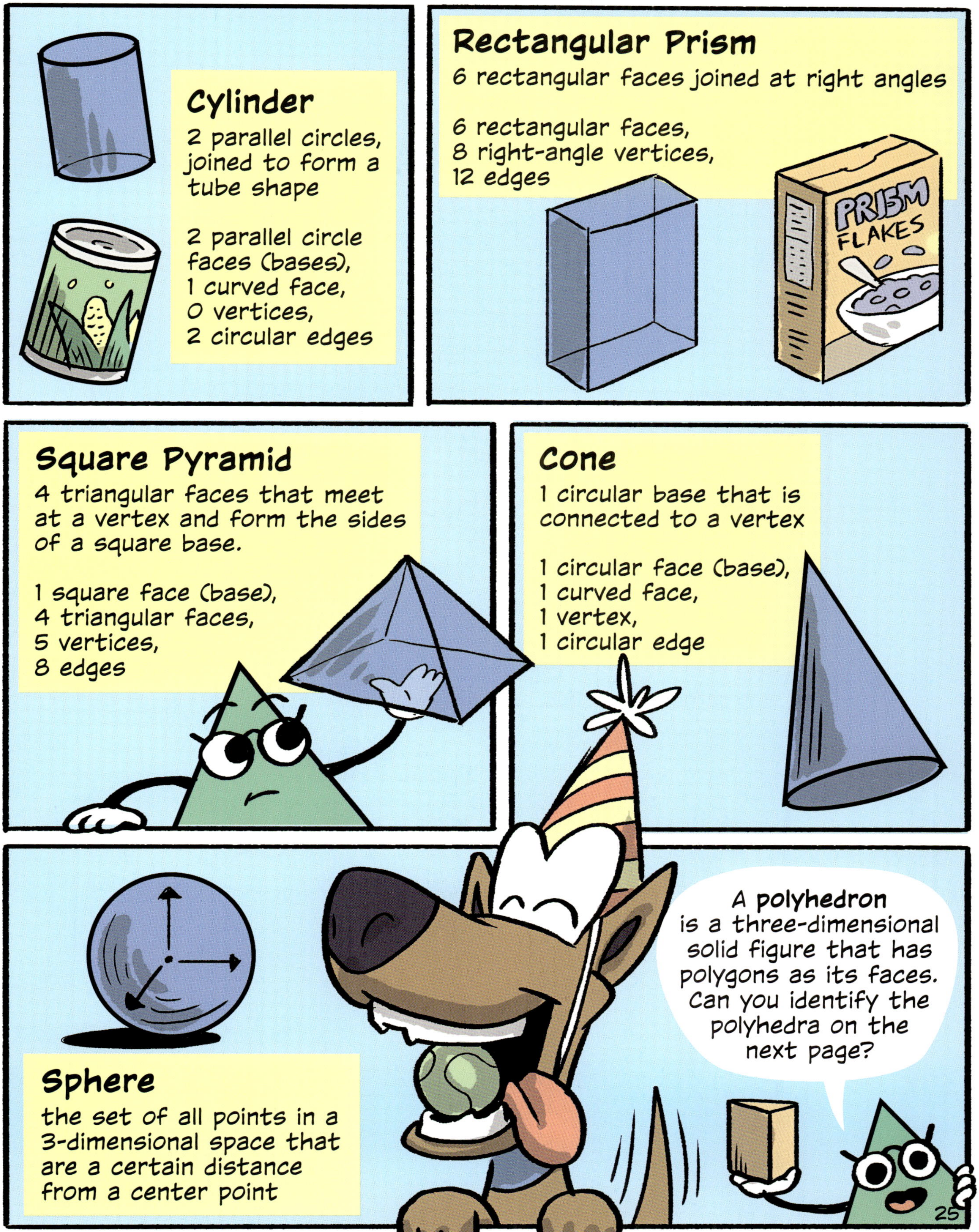
Cylinder
2 parallel circles, joined to form a tube shape
2 parallel circle faces (bases),
1 curved face,
0 vertices,
2 circular edges
Rectangular Prism
6 rectangular faces joined at right angles
6 rectangular faces,
8 right-angle vertices,
12 edges
PRISM FLAKES
Square Pyramid
4 triangular faces that meet at a vertex and form the sides of a square base.
1 square face (base),
4 triangular faces,
5 vertices,
8 edges
Cone
1 circular base that is connected to a vertex
1 circular face (base),
1 curved face,
1 vertex,
1 circular edge
Sphere
the set of all points in a 3-dimensional space that are a certain distance from a center point
A polyhedron is a three-dimensional solid figure that has polygons as its faces. Can you identify the polyhedra on the next page?

This is interesting.
I can unfold a three-dimensional figure to make a two-dimensional figure.
You're making a net.
A **net** is a two-dimensional pattern that shows the faces of a 3-D figure. You can cut apart and unfold a 3-D figure to make a net or fold a net to make a 3-D figure.
SNIFF

Hey, check this out. These figures have **symmetry**, which means their parts repeat in a pattern. I can separate each of these 3-D figures into two identical halves.

SYMMETRY
Do I have any symmetry?
Yes. I'll show you.
SYMMETRY FUN HOUSE
I can fold you across a vertical line...
or a horizontal line...
or either diagonal.
Lines of Symmetry
A **line of symmetry** divides a symmetrical figure into two halves, or mirror images. You can fold a shape across a line of symmetry to match the two halves.

For a regular polygon, the number of lines of symmetry equals the number of sides...
... but an irregular pentagon has fewer lines of symmetry.
Do you have lines of symmetry?
Why, yes, I do. Any line through the center of a circle divides it into two halves.
So, I have an infinite number of lines of symmetry.
Wow! That's infinitely cool!

CONGRUENT FIGURES
Who's up for some ice skating?
CONGRUENCY SKATING RINK
My partner and I are congruent triangles.

Two **congruent figures** have the same shape and size, but their positions may be different.
You can form congruent figures with turns, slides, and flips—in any combination.

How can you tell if two figures are congruent?
The hash marks show congruent sides, and the angle arcs show congruent angles. A pair of congruent figures must have matching sides and angles.

This parallelogram has congruent sides with the other two parallelograms. But its angles are different. So it's not congruent to them.
I'll have to try out for the solo competitions.

SIMILAR FIGURES
Hey, what happened? What does *similarity* mean?
Similar figures have the same shape, even if one shape is a mirror reflection of the other or even if the sides are different lengths.
Matching angles between similar figures have equal measurements.
Similarity
SKATING RINK

The side lengths of similar figures are **proportional**, meaning that corresponding side lengths form the same ratio. Here, the ratio of the side lengths is always 2 to 1.
Professor Circle
6
8
10
3
4
5
$\frac{6}{3}=2$ $\frac{8}{4}=2$ $\frac{10}{5}=2$

WE DID IT!
We did it! We learned a lot of geometry and rebuilt our world!
GEOMETRY WORLD
SKATE PARK
SYMMETRY FUN HOUSE
POLYGON PARK
THEATER
CONGRUENCY
SIMILARITY SKATE PARK
Chaos, you are a fine dog. But now it's time for you to go home.
SCRATCH
SCRATCH
CRASH!

MEOW!
!
Oh well! Back to the drawing board!

SHOW WHAT YOU KNOW

1. Classify each angle as acute, obtuse, right, or straight

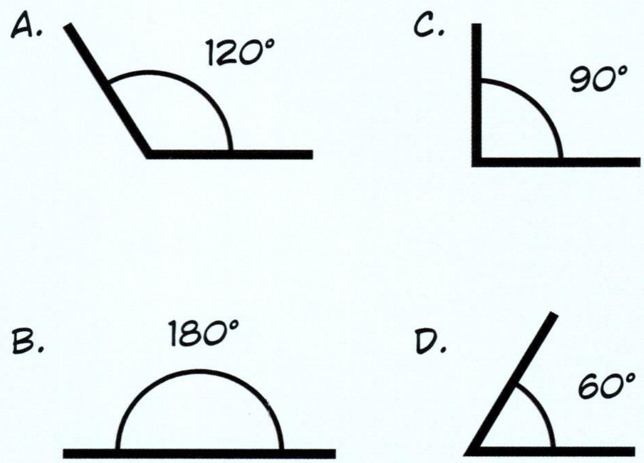

2. Answer the questions using the diagram below.

A. Line a is **parallel** to which other line?

B. Line d is **perpendicular** to which other line?

C. Line d is **parallel** to which other line?

D. True or false. Line b is **perpendicular** to line c.

E. True or false. Line c is **parallel** to line e.

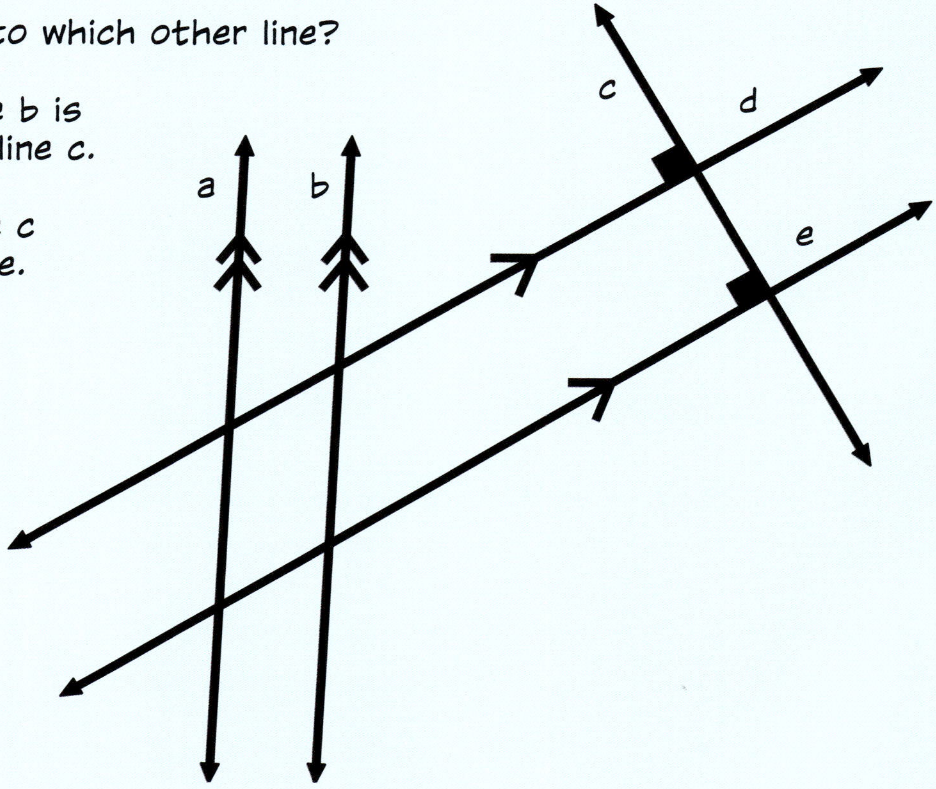

A. This triangle is an acute triangle because two of its angles are acute.

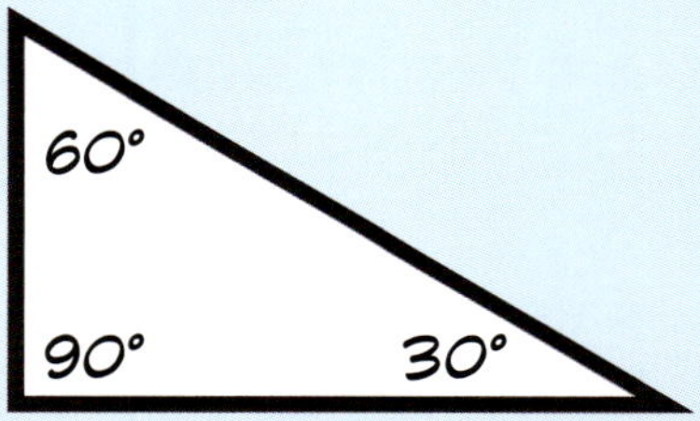

B. This triangle is an isosceles triangle because two of its sides are equal in length.

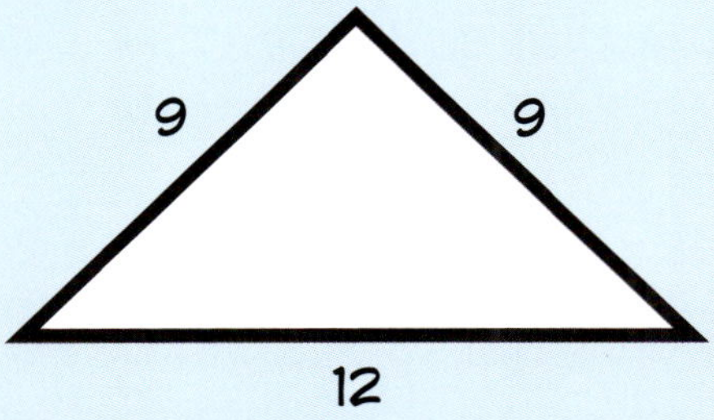

C. This triangle is an equilateral triangle because all of its sides are different in length.

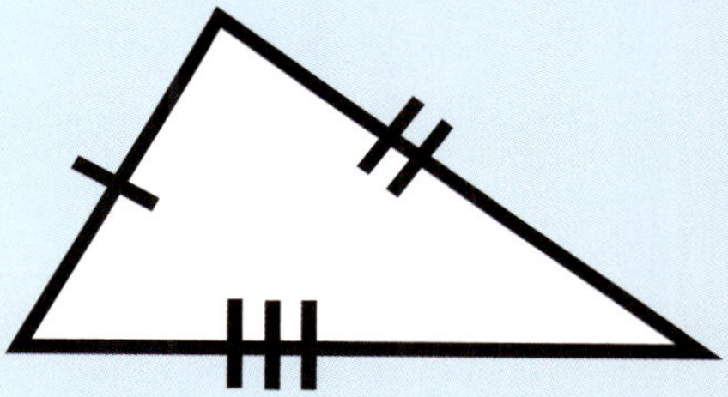

4. Multiple choice. Pick the word that correctly fills in the blank.

1. Two __________ figures have the same size and shape.
   A. congruent
   B. parallel
   C. similar

2. A __________ is a polygon with 5 sides and 5 angles.
   A. circle
   B. hexagon
   C. pentagon

3. A __________ is a quadrilateral with 4 sides of equal length.
   A. rectangle
   B. rhombus
   C. trapezoid

4. A __________ is a three-dimensional figure with 1 square face and 4 triangular faces.
   A. cone
   B. cylinder
   C. square pyramid

5. A __________ has one endpoint and goes on forever away from the endpoint.
   A. line
   B. ray
   C. segment

See page 40 for answers.

ANSWERS

1. A. Obtuse Angle - the angle measures between 90° and 180°.
B. Straight Angle - the angle measures exactly 180°.
C. Right Angle - the angle measures exactly 90°.
D. Acute Angle - the angle measures between 0° and 90°.

2. A. Line a is **parallel** to line b because they both have two arrow marks.
B. Line d is **perpendicular** to line c because they form a right angle.
C. Line d is **parallel** to line e because they both have one arrow mark.
D. False. Line b is not **perpendicular** to line c because they do not form a right angle.
E. False. Line c is not **parallel** to line e. The 2 lines are **perpendicular** because they meet to form a right angle.

3. A. False - this is a right triangle because one of the angles is 90°.
B. True
C. False - this is a scalene triangle because all of its sides are different in length.

4. **Multiple choice:**
1. Two congruent (A) figures have the same size and shape.
2. A pentagon (C) is a polygon with 5 sides and 5 angles.
3. A rhombus (B) is a quadrilateral with four sides of equal length.
4. A square pyramid (C) is a three-dimensional figure with 1 square face and 4 triangular faces.
5. A ray (B) has one endpoint and goes on forever away from the endpoint.